A Day of Adventure:
PUSHES AND PULLS

Written by April Smith
Illustrated by Dalila V. Smith

Simplify Science®

Text copyright © 2024 Performing in Education, LLC

Illustrations copyright © 2023 Dalila V. Smith

All rights reserved. No part of this publication may be reproduced, distributed, or transmitted in any form or by any means, including photocopying, recording, or other electronic or mechanical methods, without the prior written permission of the publisher, except in the case of brief quotations embodied in critical reviews and certain other noncommercial uses permitted by copyright law. For permission requests, write to the publisher at the address below.

Performing in Education LLC
help@performingineducation.com
500 N. Estrella Parkway #B2 #496
Goodyear, AZ 85338

First Edition, 2024

ISBN: 979-8-89217-109-0

Typeset in Gill Sans

Special thanks to Rebecca Riddle, Sarah Wilson, and Lydia Pearson for their contribution to this series.

Visit us at SimplifyScience.com for our companion curriculum.

A Note For Teachers & Home Educators:

After searching for vocabulary-rich picture books for the science standards and coming up empty-handed, we decided to create this our own. We hope you find this book to be a valuable resource. Please share your experience with our science picture books by emailing us at **help@simplifyscience.com**.

View our other titles at **simplifyscience.com/books**.

Find hands-on lessons for the science standards at
simplifyscience.com.

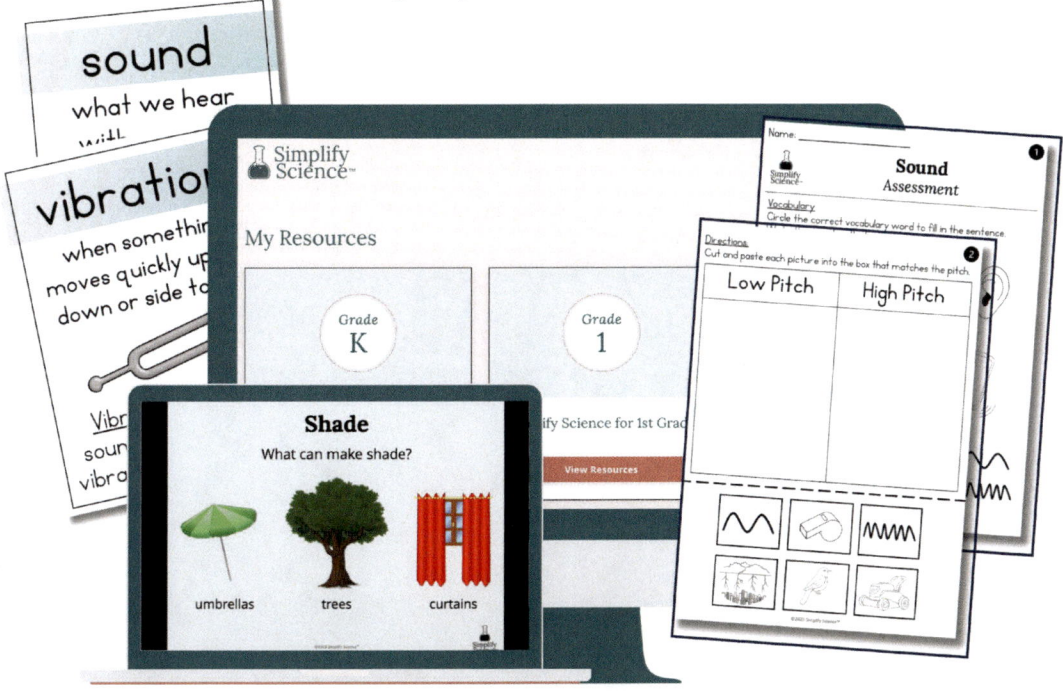

Join Mia on a day full of adventures!
Can you spot the pushes and pulls in her exciting world?

Pg 1

Remember that a **push** moves things away from you, like when using a lawnmower. A **pull** moves things closer to you, like when you fly a kite. Let's get started!

Good morning, Mia! What did Mia push to let the sunshine in?

Breakfast time! When Mia opened the fridge, was it a push or a pull?

Off to school! Mia pressed a button to call the elevator. Was it a push or a pull?

Pg 8

At recess, Mia and her friends played on the playground. Can you identify the pushes and pulls in their play?

Time for lunch! Mia opened her backpack to get her lunch. Was it a push or a pull?

Pg 13

Back to class! Mia walked into the classroom. What did she do to the door?

After school, Mia visits her friend Ellie's house. Do you see any pushes and pulls in Ellie's neighborhood?

Time to head home! Mia crossed the street with her backpack. What pushes and pulls do you notice?

Pg 18

Time for dinner! Mia got ready to eat. Did she push or pull her chair?

After dinner Mia played with a toy car.

Fast and Slow:

Here's the exciting part: a hard push makes things move faster! Mia gave a gentle push to the toy car, and it moved slowly. When she pushed harder, it zoomed!
The strength of the push affected the **speed**.

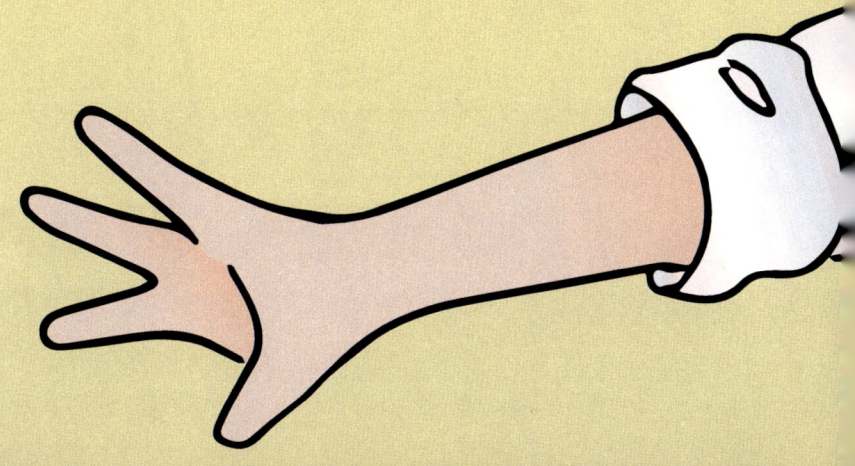

Well done! You followed Mia's day of adventures and learned about pushes, pulls, and speed. Can you think of other pushes and pulls in your own day?

Glossary

Pull: Brings things closer to you

Push: Moves things away from you

Speed: How fast or slow something moves

Made in the USA
Las Vegas, NV
11 October 2024